Hazardous materials in the building industry

Siegmund Kaub

Hazardous materials in the building industry

Knowledge base for architects and real estate experts

Springer

Siegmund Kaub
Kaub Umwelt Consult
Roßdorf, Germany

ISBN 978-3-658-49149-9 ISBN 978-3-658-49150-5 (eBook)
https://doi.org/10.1007/978-3-658-49150-5

Translation from the German language edition: "Schadstoffe im Bauwesen" by Siegmund Kaub, © Der/die Herausgeber bzw. der/die Autor(en), exklusiv lizenziert an Springer Fachmedien Wiesbaden GmbH, ein Teil von Springer Nature 2024. Published by Springer Fachmedien Wiesbaden. All Rights Reserved.

This book is a translation of the original German edition "Schadstoffe im Bauwesen," 4th edition, by Siegmund Kaub, published by Springer Fachmedien Wiesbaden GmbH in 2024. The translation was done with the help of an artificial intelligence machine translation tool. A subsequent human revision was done primarily in terms of content, so that the book will read stylistically differently from a conventional translation. Springer Nature works continuously to further the development of tools for the production of books and on the related technologies to support the authors.

This Springer imprint is published by the registered company Springer Fachmedien Wiesbaden GmbH, part of Springer Nature.
The registered company address is: Abraham-Lincoln-Str. 46, 65189 Wiesbaden, Germany

Foreword

Hazardous construction materials can be of significant economic importance in building and real estate industries.

Examples of this include the assessment of potential contaminated sites, the remediation of construction pollutants or the costly disposal of environ-mentally relevant construction waste.

This book is a useful guide for all industry stakeholders: clients and investors, project developers, construction supervisors and project managers, architects and planners, general contractors and construction companies.

It covers important vocabulary used in contaminated site remediation, waste management, building pollutants, occupational safety and toxicology, environmental analysis and environmental geology.

Its lexical structure and easy-to-understand explanations of relevant terminology tailored to its target groups and their professional practice, provide the reader with a broad yet comprehensive initiation into an increasingly complex subject matter.

Roßdorf, Germany Siegmund Kaub

Contents

A

Activated Carbon Activated carbon is a solid substance primarily composed of carbon. It is characterized by its large internal surface area (up to 2000 square meters per gram) and the resulting high porosity.

The raw materials used for the production of activated carbons are carbon-containing resources such as wood, peat, lignite, coal, or coconut shells. In the practice of dealing with contaminated sites, activated carbons are particularly used as adsorbents for the remediation of soil air and groundwater damage.

Aliphatic Hydrocarbons Organic compounds whose carbon atoms are arranged in straight or branched chains. Aliphatic hydrocarbons are found in the following petroleum products: fuels (heating oils), motor fuels (gasoline, diesel, kerosene), lubricants (lubricating oil), or solvents (petroleum ether).

S. Kaub, *Hazardous Materials in the Building Industry*,
https://doi.org/10.1007/978-3-658-49150-5_1

In the context of the remediation of contaminated sites such as former gas station sites, aliphatic hydrocarbons are often detected as contaminants in environmental media such as soil, soil air, and groundwater.

Adsorption The accumulation of gaseous or dissolved substances on a carrier material with a large specific surface area due to intermolecular van der Waals forces.

Adsorption processes are particularly used in the remediation of contaminated sites.

Examples include the remediation of environmental media such as soil air and groundwater from volatile pollutants like chlorinated hydrocarbons (CHCs) or BTEX using adsorbents such as ▶ Activated carbon.

Aerobic Term for the lifestyle of animal or plant organisms that require oxygen for respiration.

Aerobic microorganisms are used in the context of ▶ biological soil remediation for the targeted degradation of soil pollutants, such as mineral oils or other organic compounds.

Anaerobic Term for the lifestyle of animal or plant organisms that do not require oxygen for respiration.

Anaerobic microorganisms are used in the context of biological soil remediation for the targeted degradation of soil pollutants, such as mineral oils or other organic compounds.

Anthropogenic In the practice of contaminated sites, the term "anthropogenic" is used to describe the human-caused introduction of pollutants into the environment.

Typical anthropogenic burdens are the fill materials encountered in many inner-city urban centers.

The fill materials generated during construction activities are often contaminated with environmentally relevant pollutants such as heavy metals or hydrocarbons.

As contaminated excavated soil, fill materials cannot be utilized without restrictions and must therefore be disposed of at a comparatively high cost.

Aquifer In hydrogeology, an aquifer is understood to be a rock body in solid or loose rock that contains ▶ groundwater and is hydraulically suitable for transmitting it.

The term aquifer is often used synonymously with the term groundwater reservoir.

Depending on the type of flow-effective cavities, a fundamental distinction is made between pore and fracture groundwater reservoirs.

Aquifers are important resources, particularly for the supply of population with drinking water in water protection areas.

Interventions in the groundwater system, for example, within the scope of water management for construction projects, generally require official approval.

Aromatic Hydrocarbons Aromatic hydrocarbons are ring-shaped organic compounds with the basic building block ▶ Benzene.

Aromatic hydrocarbons are produced, for example, during fire damage (dioxins, furans), are additives in fuels (benzene, toluene, xylene), found in tar-containing road debris or parquet adhesive (PAHs), or serve as plasticizers in joint sealants (PCBs).

Aromatic hydrocarbons are particularly problematic for humans because some of them are classified as ▶ carcinogenic or mutagenic.

Asbestos Asbestos is a collective term for a group of six naturally occurring mineral raw materials, of which chrysotile (white asbestos) has the greatest economic significance.

Due to technologically desirable properties such as non-flammability, elasticity, adsorption capacity, etc., asbestos was extensively used in the construction industry, particularly in the field of fire protection.

The health hazards posed by asbestos are primarily due to the inhalation of microscopically small, respirable fibers, which can lead to ▶ asbestosis, lung cancer, or mesothelioma (cancer of the pleura or peritoneum).

Depending on the mass ratio of asbestos fibers to binding agents and thus the degree of fiber binding, remediation technology distinguishes between strongly bound asbestos products (asbestos cement products) such as roofing or facade cladding and weakly bound asbestos products such as fire protection panels or sprayed asbestos.

Asbestos Remediation Asbestos remediation refers to the application of technical procedures to eliminate health hazards to building occupants caused by the construction pollutant ▶ asbestos.

Three different remediation methods are applied in the remediation of asbestos products:

Removal: Complete removal of the asbestos product
Coating: Coating the asbestos product with a plasto-elastic medium.
Spatial Separation: Dust-tight encapsulation of the asbestos product from the room

The assessment of the necessity for asbestos remediation is carried out by asbestos experts through evaluation of the hazard potential present in each individual case.

The choice of remediation method in each case depends, in addition to criteria such as the strength of the asbestos product or potential impairments due to future use, particularly on the acceptance of the affected building users.

The execution of an asbestos remediation is planned, monitored, and finally documented by expert consultants through the creation of remediation concepts. The success of a remediation measure is demonstrated, among other things, by ▶ indoor air measurement.

Asbestosis Asbestosis is a disease caused by the inhalation of asbestos fibers, resulting in pathological scarring of the lung tissue.

Asbestosis leads to a persistent cough, shortness of breath, and sputum, but often also to asbestos-related lung cancer.

The choice of remediation method [illegible] case depends, in addition to criteria [illegible] the asbestos product or potential [illegible] use, particularly on the acceptance [illegible] building users.

The execution of an asbestos [illegible] monitored and finally documented [illegible] through the execution of remediation [illegible] of a remediation measure is [illegible] things, by indoor air measurements.

Asbestosis: Asbestosis is a disease [illegible] inhalation of asbestos fibers resulting in [illegible] of the lung tissue.

Asbestosis leads to [illegible] shortness of breath and coughing, but often [illegible] lung cancer.

B

Background Concentration The pollutant concentration in soil, which consists of the natural (geogenic) baseline concentration of the soil and the ubiquitous distribution of substances as a result of diffuse inputs into the soil.

Benzene Benzene is the basic structure of aromatic hydrocarbons. Benzene is obtained during petroleum refining and is a starting material for the production of a variety of chemical products such as paints, plastics, or pharmaceuticals.

Benzene is toxicologically classified as ▶ carcinogenic and ▶ mutagenic.

In the construction industry, benzene and other aromatic hydrocarbons play an important role as pollutants in the remediation of contaminated sites (e.g., former gasworks sites) or buildings (e.g., PCB remediation).

S. Kaub, *Hazardous Materials in the Building Industry*,
https://doi.org/10.1007/978-3-658-49150-5_2

Benzo-A-Pyrene Benzo-a-pyrene (BaP) is the most important representative and key parameter of the class of Polycyclic Aromatic Hydrocarbons (▶ PAH).

Benzo-a-pyrene is formed during incomplete combustion processes and is therefore found in tobacco smoke as well as in vehicle exhaust gases or coke oven emissions.

Benzo-a-pyrene is toxicologically classified as carcinogenic and mutagenic.

In the construction industry, benzo-a-pyrene plays an important role as a pollutant in the remediation of contaminated sites such as former gasworks or coke ovens, as well as in building renovations (tar-containing parquet floor adhesives).

Biological Soil Remediation Biological soil remediation refers to all processes for the remediation of contaminated soils using pollutant-degrading microorganisms.

Biological soil remediation is primarily used for the remediation of contaminated sites, but also for the cleaning of contaminated soils generated during construction projects.

The methods of biological soil remediation are limited to degradable organic pollutants, such as mineral oils, and are therefore not successfully applicable to all soil contaminations.

Through various technical measures, such as mixing, aeration, moistening, or targeted changes in nutrient supply, the living conditions of microorganisms can be improved, thereby enhancing their degradation performance.

Depending on the specific circumstances, biological soil remediation can occur in the natural storage environment of the soil (in situ), or—after excavation—on the construction site (on site) or in specialized soil treatment centers outside the construction site (off site).

Black Area A black area refers to the area contaminated by pollutants during work in contaminated zones.

The black area (remediation area) is typically separated from the public, uncontaminated area ("white area") by special barriers to prevent the spread of pollutants.

Access to the black area is generally restricted to authorized personnel under appropriate occupational safety conditions (e.g. protective suits, respiratory protection).

Black-and-White Facility When working in contaminated areas, a black-and-white facility serves as the interface between the remediation area (black area) and the public area (white area).

A black-and-white facility allows for the spatially separated storage, keeping, and re-donning of street and work clothing and includes the necessary sanitary facilities (washbasin, shower, toilets) for remediation personnel.

BTEX Abbreviation for the chemical compound benzene and the aromatic hydrocarbons derived from it: toluene, ethylbenzene and xylene.

BTEX aromatics were particularly used as additives for gasoline but also as industrial solvents.

Contamination of environmental media by aromatic hydrocarbons is found at former gas station sites as well as in metalworking operations, the production of plastics, paints and varnishes, or pesticides.

BTEX aromatics often contaminate not only soil and groundwater but also soil air, which then must be remediated through extraction and subsequent purification ▶ soil air remediation, a process that is time-consuming and costly.

Building Contaminants Building contaminants are substances that pose a health risk to humans within buildings.

Building contaminants include construction materials with harmful substances such as asbestos, wood preservatives, or PCBs. Additionally, furnishings like furniture (formaldehyde), floor coverings (PAHs), or wallpapers (solvents) can contribute to indoor pollution.

Before the remediation of building contaminants, expert assessors analyze potential sources of contamination both qualitatively and quantitatively and, if necessary, develop a case-specific remediation plan.

For certain building contaminants, such as PCBs, a trial remediation is conducted in representative areas before extensive remediation measures are implemented to test the effectiveness of the chosen remediation method.

The remediation of building contaminants is generally labor-intensive, time-consuming, and therefore costly. In addition to the potentially very high remediation costs, the need to secure alternative accommodations for the duration of the building remediation may also be necessary.

C

Carcinogen Carcinogens are generally defined as factors that can lead to the development of cancer in humans or animals.

In the construction industry, carcinogenic substances are particularly encountered during the remediation of hazardous waste sites or building contaminants.

Carcinogenicity Index The Carcinogenicity Index (CI) is a dimensionless parameter used to evaluate the cancer-causing potential of fibers with specific geometric dimensions (L > 5 µm, D < 3 µm, L/D ratio > 3), also referred to as > WHO fibers.

The determination of the CI value is particularly used for the assessment and classification of unknown mineral wool products. The CI value is calculated based on the mineral composition of the fibers.

S. Kaub, *Hazardous Materials in the Building Industry*,
https://doi.org/10.1007/978-3-658-49150-5_3

Chlorinated Hydrocarbons Chlorinated hydrocarbons (CHCs) are hydrocarbons in which one or more hydrogen atoms have been replaced by the element chlorine.

Among the most well-known representatives of chlorinated hydrocarbons are the compounds trichloroethene (Tri) and tetrachloroethene (Per), which were used as industrial solvents, particularly in chemical cleaning and metalworking operations.

Chlorinated hydrocarbons often appear as contaminants in environmental media such as soil, soil air, and groundwater during the remediation of contaminated sites.

The remediation of CHC-related environmental damage is often prolonged and therefore costly, especially in cases of groundwater contamination.

Chlorofluorocarbons Chlorofluorocarbons (CFCs) are hydrocarbons in which one or more hydrogen atoms have been replaced by the elements fluorine and chlorine.

CFCs are particularly significant in the construction industry during the dismantling of refrigeration and air conditioning systems during the demolition or renovation of buildings.

Construction Waste Construction waste refers to waste generated during the construction, renovation, refurbishment, or demolition of buildings.

Construction waste primarily includes mineral mass waste such as excavated soil, demolition debris, or road rubble, construction site waste like packaging materials, or hazardous construction waste such as asbestos or contaminated excavated soil.

Responsible for the proper ▶ disposal of waste generated during construction projects are both the clients and the executing companies in their roles as waste producers or waste owners.

Containment In the context of contaminated site remediation, containment measures prevent or reduce the long-term spread of pollutants without removing the pollutants themselves.

Containment measures are only permissible for the remediation of a contaminated site if they ensure that the pollutants remaining in the soil or the contaminated site do not pose any long-term dangers, significant disadvantages or considerable nuisances to individuals or the public.

Examples of containment measures include sealing or encapsulation of contaminated sites. Additionally, covering contaminated soils or sites with suitable soil layers or sealing can also be considered as containment measures.

When remediating contaminated sites, the party responsible for remediation should always assess whether containment measures or the removal of pollutants are the most cost-effective long-term remediation strategy.

Contaminated Area A contaminated area is generally understood to be a region that contains hazardous substances in quantities harmful to humans and/or the environment.

Contaminated areas particularly include polluted properties ▶ contaminates sites, production sites, or buildings that are burdened with construction pollutants such as asbestos or PCBs.

In the context of remediation measures, the area contaminated by hazardous substances is also referred to as the remediation area or "black area."

Contaminated Sites Contaminated sites refer to: 1. Decommissioned waste disposal facilities and other properties where waste has been treated, stored, or deposited (old deposits) and 2. Properties of decommissioned facilities and other properties where environmentally hazardous substances have been handled (old sites), which cause harmful soil changes or other dangers to individuals or the public.

Whether a property is considered a contaminated site is determined by the responsible contaminated site authority based on an expert risk assessment on a case-by-case basis.

Once a contaminated site is identified, the relevant parcels are recorded in an officially maintained contaminated site register.

If the party responsible for remediation provides evidence to the competent authority of the successful remediation of the affected parcel, the contaminated site designation is lifted by the authority, and the register entry is deleted.

Developers should note that regardless of an official contaminated site designation, significant volumes of filled soils may arise during construction projects as part of earthworks, which, due to their pollutant content, must be disposed of as contaminated excavated soil at a relatively high cost.

Contaminated Site Management The processing of contaminated site projects generally occurs in three interdependent steps: identification, risk assessment, and remediation.

Each phase concludes with a decision by the responsible contaminated site authority on whether further investigations or remediation measures are necessary for a given area, or if the area is excluded from further processing due to a lack of concrete suspicion.

If contaminated sites are identified in advance of construction activities, the type, scope, and cost of necessary remediation measures can be determined early by expert assessors.

This provides builders and project managers with important decision-making foundations for construction scheduling and cost control of the overall project.

Contaminated Site Register The authorities responsible for contaminated sites maintain a computer-based data collection for their regional areas of responsibility, containing all available information (parcels, pollutants, risk potential, etc.) about contaminated sites or suspected contaminated areas.

The contaminated site or suspected area registers can be reviewed by builders, for example, in advance of property purchases or planned construction activities at the responsible contaminated site authority.

Contaminated Site Remediation Contaminated site remediation refers to the application of technical procedures with the primary goal of averting the dangers posed by a contaminated site to human life and health, considering the existing or planned use of a property.

Two groups of procedures are distinguished for the remediation of contaminated sites:

Decontamination procedures: Methods for the removal or reduction of pollutants.

Containment procedures: Methods that prevent or reduce the spread of pollutants in the long term without removing them.

The procedures applied in the remediation of a contaminated site in each case depend on many factors, of which the risk potential posed by the contaminated site, the geological and hydrogeological site conditions, and not least the available financial resources for damage remediation are the most important.

Contamination The term contamination, as well as the often synonymously used terms "pollution" and "impurity," are used in various areas of environmental engineering.

In the practice of dealing with contaminated sites, the term contamination is used to describe the degree of pollution of soil, soil air, or groundwater following the introduction of pollutants by humans.

Property owners should be aware that from a certain level of contamination, remediation measures are generally required, which can significantly influence the cost and process of a construction project depending on the type and extent of the impurities.

In waste management, the term contamination refers to the pollution of a type of waste, e.g., excavated soil, construction debris, or demolition wood, with environmentally relevant pollutants.

For property owners, it is important to note that the permissible disposal route for construction waste and thus the disposal costs significantly depend on the type and extent of any existing contamination.

Conversion In the context of contaminated sites, conversion refers to the civilian repurposing of formerly military-used properties.

Developers and investors interested in the conversion of these properties must consider that, in addition to potential soil contamination, the building materials of existing structures are often significantly contaminated with hazardous substances, leading to increased costs for demolition and ▶ disposal of demolition materials.

Cyanides Cyanides are the salts of hydrocyanic acid (HCN).

Of particular importance to humans and the environment are the highly water-soluble and very toxic alkali and alkaline earth cyanides.

Cyanides are particularly significant for the construction industry in the remediation of ▶ contaminates sites such as electroplating facilities or hardening plants.

D

Danger General police and regulatory law defines danger as a situation that, if events proceed unhindered, is likely to result in damage to public safety and order or their protected assets in the foreseeable future.

The presence of a danger to individuals or the public is a crucial criterion for the responsible contaminated site authority in deciding whether an area suspected of contamination is a ▶ contaminated site or not.

Declaration Analysis In environmental analytics, a declaration analysis refers to a chemical examination of a ▶ waste to clarify the disposal route, focusing on various parameters (pollutants).

In the construction industry, declaration analyses are crucial when contaminated construction waste, such as polluted excavated soil or contaminated construction and demolition wood, needs to be disposed of during construction projects.

S. Kaub, *Hazardous Materials in the Building Industry*,
https://doi.org/10.1007/978-3-658-49150-5_4

Decontamination In environmental engineering, decontamination refers to the removal or reduction of pollution caused by harmful substances.

Decontamination methods are particularly used for the remediation of ▶ contaminated site as well as for the remediation of contaminated soils generated during construction projects or for the remediation of building pollutants.

Examples of decontamination methods in the context of contaminated site remediation include the excavation and treatment of contaminated soils or the remediation of environmental media such as soil air and groundwater.

The removal of building pollutants such as asbestos, PCBs, or PAHs is also referred to as building decontamination.

Demolition The demolition of structures or parts of structures is carried out depending on the demolition object using mechanical-hydraulic methods such as grabbing, striking, or pressing, thermal separation of structural components by flame cutting, or demolition using blasting techniques.

The demolition of buildings has been and still is carried out with the wrecking ball immediately after the buildings are vacated.

This approach results in increased sorting efforts for the demolition contractor or a multiplication of disposal costs for the now mixed demolition materials (construction mixed waste).

Due to decreasing landfill capacities and drastically increased disposal costs, the trend is moving away from uncontrolled demolition towards ▶ Selective demolition.

Demolition Debris Demolition debris refers to mineral material, also with minor foreign components (<5 Vol.%), generated during the construction, renovation, refurbishment, or demolition of buildings.

Depending on the previous use of the building, demolition debris may be contaminated with various pollutants such as oils, fats, or heavy metals and must be subjected to special ▶ disposal in these cases.

Derivative In chemistry, a derivative refers to compounds derived from a chemical substance, often formed by the substitution of atoms or groups of atoms in a single reaction step.

Many pollutants encountered in the remediation of contaminated sites or building pollutants, such as PAHs, PCBs, dioxins, or nitro compounds like TNT, are derivatives of ▶ benzene.

Dermal In environmental medicine, dermal refers to the absorption of pollutants through the skin.

Pollutants that can be absorbed dermally include, for example, aromatic hydrocarbons such as benzene, PAHs, and PCBs, or wood preservatives like PCP or lindane.

To prevent skin absorption of the aforementioned pollutants, workers during the remediation of contaminated sites or building pollutants are required to wear special protective suits, gloves, and boots.

Detailed Investigation A detailed investigation refers to an in-depth examination of a suspected contaminated site conducted by expert assessors.

The goal of a detailed investigation (in-depth exploration) is to determine the quantity and spatial distribution of pollutants, their mobile and mobilizable fractions, their potential spread in soil, water, and air, as well as the possibility of their uptake by humans, animals, and plants.

The report on the detailed investigation of a suspected site concludes the ▶ risk assessment as the second main phase of contaminated site management.

Based on the detailed investigation, the responsible contaminated site authority decides whether a suspected site should be classified as a contaminated site and what further measures (immediate actions, remediation investigations) are necessary.

Detection Limit In environmental analysis, the detection limit refers to the pollutant concentration at which the substance under investigation can just be detected, depending on the specific analytical method used.

If the pollutant content is below the detection limit of the analytical method, the substance in question is considered not detected (n.d.).

In contrast to the detection limit, the quantification limit describes the concentration of a substance at which it can be reliably and reproducibly measured, depending on the analytical method.

Dioxins Polychlorinated dibenzodioxins (PCDD), often simply referred to as dioxins, is a collective term for a group of 75 highly toxic organic chlorine compounds.

Of particular importance is the acutely toxic, carcinogenic, and mutagenic tetrachlorodibenzodioxin (2,3,7,8-TCDD, Seveso poison), on whose toxicity the total concentrations of the various dioxins are usually calculated.

Dioxins are formed as unwanted by-products during the production of various chlorinated organic compounds, but also during incomplete combustion processes such as the smoldering of plastics or insulated cables in waste incineration plants or the smelting of plastic-coated metal waste.

In the construction industry, dioxins play an important role in the remediation of contaminated sites, fire damage or former industrial facilities (e.g. metal smelters).

Disposal The disposal of waste serves to permanently exclude waste from the circular economy and is carried out, for example, through chemical-physical treatment, deposition in above-ground or underground landfills, or incineration in special incineration plants.

The permissible disposal method for waste is fundamentally dependent on the type of waste, considering the nature and level of any present pollutants, as well as the regulatory requirements of the intended disposal facility.

Waste generated during construction projects that needs to be disposed of includes, for example, asbestos products, heavily contaminated old wood, or excavated soil with harmful impurities.

E

Electrical Conductivity Electrical conductivity is defined as the reciprocal of electrical resistance and characterizes a material's ability to conduct electric current.

In environmental analysis, electrical conductivity is particularly used to assess potential contamination of groundwater or drinking water.

High electrical conductivities can also be of natural origin, so comparative measurements of uncontaminated water samples are recommended.

The electrical conductivity of drinking or service water ranges between 300–600 μS/cm.

Eluate In environmental analysis, eluates are aqueous solutions of soil samples prepared according to standardized procedures.

By examining soil eluates, it can be determined whether certain pollutants present in the soil, such as heavy metals,

S. Kaub, *Hazardous Materials in the Building Industry*,
https://doi.org/10.1007/978-3-658-49150-5_5

can be dissolved by precipitation or percolating water and subsequently enter the groundwater.

The examination of eluates is therefore particularly used for risk assessment of contaminated sites and to determine whether and how contaminated soil masses generated during construction activities can be utilized.

Emission Emissions refer to air pollutants, noise, vibrations, light, heat, radiation and similar phenomena emitted from industrial facilities, machinery, vehicles, or properties.

The maximum permissible emissions are regulated by national laws.

In the context of construction projects, emissions of dust and noise are particularly significant.

Encapsulation Encapsulation refers to technical barrier systems that serve as protective measures to prevent further spread of pollutants from a contaminated site or area.

Depending on the location of the barrier, encapsulations are distinguished between vertical sealing systems, such as sheet pile, narrow, or diaphragm walls, horizontal injection systems for base sealing, and generally mineral surface coverings.

Environmental Analysis Environmental analysis involves the chemical-physical determination of environmentally relevant pollutants in solid, liquid, or gaseous media.

Common investigative methods in environmental analysis include, for example, scanning electron microscopy (SEM) for determining asbestos, atomic absorption spectroscopy (AAS) for detecting heavy metals, or gas chromatography (GC) for pollutants such as PAHs, PCBs, or dioxins.

In the field of contaminated site remediation, chemical analysis of soil, soil air, and groundwater samples is particularly used to assess potential human health risks and to determine any necessary remediation measures.

Before disposing of construction waste such as excavated soil, construction debris, or demolition wood, it is generally necessary to analyze these wastes for various potential pollutants as part of a ▶ declaration analysis to verify the permissibility of a planned disposal route.

To verify the success of a building remediation, collected indoor air samples are analyzed for construction-related pollutants such as asbestos or PCBs.

EPA Abbreviation for the United States Environmental Protection Agency, headquartered in Washington, D.C.

Regulations and recommendations issued by the EPA are applied in environmental analysis, for example.

An analysis value labeled as PAH according to EPA indicates that the 16 most common polycyclic aromatic hydrocarbons in the environment (▶ PAH) were examined, and subsequently, the sum of the individual concentrations was calculated.

Excavation In environmental engineering, excavation refers to the removal of contaminated soil material using construction machinery.

The excavation of pollutant-laden soils occurs during the remediation of ▶ contaminated sites as well as during construction projects where known or newly discovered foreign deposits in the soil become apparent during earthworks.

Verification of successful excavation remediation is typically achieved through sampling of the walls and floor of the excavation site and chemical analysis of the soil samples for the specific pollutants present.

Depending on the type and concentration of pollutants detected in the soil, special occupational safety measures (e.g., hydraulic excavators with filter cabins) are required during the excavation of contaminated soils.

The excavated contaminated soil masses are secured on-site in containers or with foil coverings, chemically analyzed for waste classification, and subsequently disposed of.

Exposure In environmental engineering, exposure refers to the manner in which a pollutant comes into contact with a receptor, such as people, animals, plants, soil, or groundwater.

Pollutant exposures are particularly relevant in the construction industry, especially during the remediation of contaminated sites or building pollutants like asbestos or PCBs.

To prevent health damage from the pollutants being remediated, appropriate occupational safety measures are legally required.

F

Fiber Dusts Fiber dusts are defined as dusts containing artificial inorganic mineral fibers with specific, critical dimensions (fiber length $L > 5\mu m$, fiber diameter $D < 3\mu m$, $L : D > 3$). Fiber dusts are also referred to as ► WHO fibers.

The construction industry is particularly affected by fiber dusts during the dismantling of mineral wool insulation materials used for thermal and sound insulation in the field of building construction.

For older mineral wool products there is a fundamental assumption of potential carcinogenic risk.

Filling In soil science, filling refers to the material placed above the naturally occurring soil layers of anthropogenic origin.

S. Kaub, *Hazardous Materials in the Building Industry*,
https://doi.org/10.1007/978-3-658-49150-5_6

Fill soils are a heterogeneous mixture of soil with various foreign components depending on previous land use, such as remnants of construction debris, bricks, glass, ash, or slag.

Fill soils are primarily found in inner-city urban centers and range in thickness from a few decimeters to several meters.

Fill soils are particularly significant for builders because they often contain environmentally relevant pollutants such as ► PAH or ► heavy metals and therefore incur increased disposal costs for contaminated excavated material during earthworks.

Formaldehyde Formaldehyde is a colorless gas with a pungent odor, for which there is a substantiated suspicion of carcinogenic potential.

Formaldehyde is environmentally significant for humans as an indoor pollutant, where it is used as an adhesive additive in furniture and fixtures made of particleboard.

Other possible sources of formaldehyde include paints, varnishes, parquet sealants, carpet adhesives, and also tobacco smoke.

In cases of mucous membrane irritation in the nose, eyes, and throat or unexplained allergies, potential sources of formaldehyde should be considered, and if necessary, the indoor air should be checked through appropriate measurements.

Former Site Former sites are properties of decommissioned facilities and other properties where environmentally hazardous substances have been handled.

Former sites include, for example, former industrial operations such as foundries or gasworks, as well as decommissioned chemical cleaning facilities or gas station sites.

Important information about former sites (pollutant inventory, spatial extent, parcels, etc.) is available to builders through consultation of officially maintained contaminated site or suspected area registers.

Furans Polychlorinated dibenzofurans (PCDF), often simply referred to as furans, is a collective term for a group of 135 highly toxic organic chlorine compounds.

The chemical structure, occurrence, environmental behavior, and significance of furans for the construction industry are comparable to those of ► dioxins.

G

Geogenic Refers to the concentration of pollutants in the soil that results from the naturally occurring baseline concentration.

Typical geogenic soil pollutants include heavy metals such as lead or arsenic.

In contrast to geogenic soil contamination is the introduction of pollutants by humans, known as anthropogenic soil contamination.

For property owners, it is important to note that geogenic pollutant levels alone generally do not constitute a ▶ contaminated sites and therefore do not typically require remediation.

Groundwater Subsurface water that fills the pores, fractures, or karst cavities within rock formations and whose movement is solely determined by gravity.

Groundwater is a critical resource, particularly in the context of remediating contaminated sites.

S. Kaub, *Hazardous Materials in the Building Industry*,
https://doi.org/10.1007/978-3-658-49150-5_7

The typically required lowering of the groundwater level and the discharge of extracted groundwater during major construction projects require official approval.

Builders must ensure that any contaminated groundwater encountered during construction dewatering is treated before discharge to comply with local discharge limits.

Groundwater Remediation Groundwater remediation encompasses all technical methods for removing groundwater contaminants (active methods) or preventing further spread of pollutants that have entered the groundwater (passive methods).

Active remediation methods (pump and treat methods) are characterized by the extraction of contaminated groundwater with energy input, followed by treatment through processes such as adsorption, stripping, or ion exchange.

In passive methods (hydraulic containment measures), the contaminated groundwater flows without energy input through a wall constructed in the aquifer, containing reactive or adsorptive materials for pollutant removal.

The choice of remediation methods for a specific groundwater contamination case largely depends on the spectrum of pollutants present and the geological and hydrogeological conditions of the project.

The construction industry encounters groundwater remediation particularly in the preliminary stages of construction projects (contaminated site remediation) or in the initial phase of construction execution (construction dewatering).

Groundwater Table Depth The vertical distance between a point on the earth's surface and the groundwater surface of the uppermost aquifer.

The groundwater table depth plays a crucial role in the ► risk assessment of contaminated sites. A shallow groundwater table depth generally indicates a relatively high-risk potential from leachable soil pollutants.

H

Halogenated Hydrocarbons Halogenated hydrocarbons (HKW) are ▶ hydrocarbons in which the hydrogen atoms are partially or completely replaced by the elements fluorine, chlorine, bromine, or iodine (halogens).

The improper use of volatile halogenated hydrocarbons (LHKW) such as trichloroethylene (Tri) or tetrachloroethylene (Per) as industrial solvents in metal processing facilities and dry-cleaning operations has often led to environmental damage requiring remediation of soil, soil air, and groundwater.

Less volatile halogenated hydrocarbons (SHKW), used as wood preservatives (lindane, DDT) or as plasticizers in joint sealants and flame retardants (PCB) are now often the cause of costly building renovations.

S. Kaub, *Hazardous Materials in the Building Industry*,
https://doi.org/10.1007/978-3-658-49150-5_8

Harmful Contamination Harmfully contaminated waste refers to ▶ waste with particularly high concentrations of pollutants.

The disposal of harmfully contaminated waste requires special monitoring.

Examples of harmfully contaminated construction waste include weakly bound asbestos products, particularly contaminated old wood, or construction debris and excavated soil with harmful contamination.

Hazard Potential In the context of contaminated sites, hazard potential refers to the sum of possible risks and damages to humans or the environment that can be expected under certain conditions from a contaminated site.

The hazard potential of a contaminated site fundamentally depends on the specific inventory of pollutants and the possible pathways through which a pollutant can reach the protected target.

Hazardous Substance Hazardous substances are materials that are, for example, explosive, toxic, carcinogenic or exhibit other hazardous characteristics.

Hazardous substances are particularly encountered in the construction industry during the remediation of contaminated sites or building contaminants like asbestos, PCBs, or PAHs.

When handling hazardous substances, special measures for the occupational safety of employees must be observed. These include, in particular, the assessment of the specific hazard present in each case and the establishment of organizational, technical, and personal protective measures.

Hazardous Waste In waste management, the term hazardous waste is often used for waste that, due to its nature, quantity, or characteristics, poses significant health, air, or water hazards, is explosive or flammable, or can contain or produce pathogens of communicable diseases.

Hazardous waste is often colloquially referred to as "special waste."

Hazardous waste cannot be disposed of together with household waste in a municipal landfill but requires separate disposal in special hazardous waste landfills (SHWL) or hazardous waste incineration plants (HWIP).

Examples of hazardous waste generated during construction projects include asbestos products, PCB-containing waste, demolition wood treated with wood preservatives, or contaminated excavated soil.

HBCD Abbreviation for the halogenated hydrocarbon compound Hexabromocyclododecane. HBCD has been and is used in the construction industry for fire protection as a flame retardant in external thermal insulation composite systems such as EPS or XPS (expanded or extruded polystyrene).

Heavy Metals Heavy metals are defined as all metals with a density greater than 5 g/ccm.

Heavy metals are partly essential trace elements for humans (e.g. copper, zinc, or manganese) and partly toxic environmental pollutants (e.g. mercury, lead, or cadmium).

For the construction industry, heavy metals are primarily significant as pollutants in the ▶ filling where they lead to increased disposal costs during earthworks.

In addition to soil contamination, heavy metals often also contaminate the building substance, especially at commercially or industrially used sites such as foundries or metal smelters, thereby necessitating increased occupational safety requirements during building demolition and increased costs for the disposal of contaminated demolition materials.

Historical Investigation A historical investigation refers to the site-specific collection, preparation, and evaluation of data, facts, and findings available from written sources, maps, and aerial photographs, as well as from interviews, particularly regarding the past and present use of a potentially contaminated site.

The historical investigation is the first step in assessing the risk posed by a potentially contaminated site.

In some countries the terms historical research, in-depth survey, case-by-case research, site research or site-specific survey are used synonymously.

To reduce the risk of costly construction site delays, it is generally advisable for developers to conduct a historical investigation in addition to consulting official ▶ contaminated site registers prior to a planned construction project.

Hydrocarbons Hydrocarbons are chemical compounds of carbon (C) with the element hydrogen (H).

Due to chemical properties such as the spatial arrangement of molecules, different types of chemical bonds, and the possibility of exchanging C or H atoms, there is an overwhelming variety of individual compounds.

Based on their chemical structure, hydrocarbons are classified into chain-like (aliphatic), ring-like (cyclic), and aromatic hydrocarbons, which are based on the core structure of ▶ benzene.

Hydrocarbons often appear as contaminants to be remediated in construction and renovation projects.

In the remediation of contaminated sites, these are often petroleum products, which, for example, led to contamination of gas station sites as fuels, or chlorinated hydrocarbons (CHCs), whose improper use as solvents in metal processing or dry-cleaning operations caused environmental damage.

Hydrocarbons also appear as pollutants in the remediation of fire damage cases (dioxins/furans), joint sealants (PCBs), parquet floor adhesives (PAHs), or as wood preservatives (PCP, lindane).

I

Immission Immissions refer to the air pollutants, noise, vibrations, light, heat, radiation, and similar environmental impacts affecting humans, animals, plants, soil, water, the atmosphere, as well as cultural and other material assets.

Immobilization Immobilization refers to technical processes aimed at reducing the mobility and mobilizability of soil contaminants for emission processes such as leaching (elution), dispersion, or gas formation.

The immobilization processes serve as typical safeguarding measures primarily for the remediation of ► contaminated site.

The stabilization of soil contaminants is achieved by adding water with binding agents such as cement, bentonite, water glass, or synthetic resins.

S. Kaub, *Hazardous Materials in the Building Industry*,
https://doi.org/10.1007/978-3-658-49150-5_9

Depending on the existing soil conditions, immobilization can occur either after excavation of the contaminated soil on site or directly by injecting the stabilizing agents into the contaminated subsurface (in situ).

In Situ Methods In situ methods in the practice of contaminated site remediation refer to techniques for the remediation of environmental media such as soil, soil air, and groundwater while maintaining natural storage conditions without excavating soil material.

Typical in situ methods include soil remediation through the addition of stabilizing agents (immobilization) or pollutant-degrading bacteria (biological soil remediation), remediation of soil air through soil vapor extraction or the remediation of groundwater contamination through ▶ stripping.

In construction projects on contaminated sites, in situ methods are often employed in advance of later excavation remediation or parallel to construction activities. In both cases, a technically and economically tailored remediation plan adapted to the conditions of the construction project is recommended.

Indoor Air Measurement Indoor air measurements are used for the environmental hygiene assessment of air in indoor spaces.

In the construction industry, indoor air measurements are particularly used in the context of the remediation of ▶ building contaminants.

For example, determining the PCB content in indoor air is used to assess the need for remediation, and measuring asbestos fiber concentration is used to verify the success of a remediation measure.

The goal of an indoor air measurement, the number and arrangement of measurement points, as well as any necessary usage simulation, are determined in advance by expert consultants as part of a measurement plan.

Inhalation In environmental medicine, inhalation refers to the uptake of pollutants through breathing.

Pollutants that can be inhaled include, for example, asbestos, solvent vapors, or pollutants bound to dust particles such as PAHs or dioxins.

To prevent the inhalation of pollutants, workers involved in the remediation of contaminated sites or building pollutants are required to use respiratory protection measures tailored to the specific case.

K

Key Parameters Key parameters refer to chemical compounds whose detection during the investigation of contaminated sites can indicate specific environmental burdens.

For example, elevated concentrations of the element boron in groundwater suggest household waste deposits, while increased sulfate levels indicate construction debris deposits.

Additionally, the term key parameter is used in environmental analysis to denote the most important individual substance within a class of compounds.

For instance, the compound benzo[a]pyrene is a key parameter of polycyclic aromatic hydrocarbons (▶ PAH) and the compound 2,3,7,8-tetrachlorodibenzo-p-dioxin (TCDD) is a key parameter of ▶ dioxins.

S. Kaub, *Hazardous Materials in the Building Industry*,
https://doi.org/10.1007/978-3-658-49150-5_10

Kieselrot Kieselrot refers to leaching residues from copper extraction or mixtures with these residues, which were particularly used as surfaces for sports fields, playgrounds, and on paths and plazas.

Kieselrot contains a variety of chemical compounds that are toxic to humans. In addition to heavy metals, sulfur, and nitrogen compounds, highly toxic chlorine compounds such as dioxins and furans are of particular concern.

L

Land Recycling Land recycling (land reactivation) refers to the conversion of unused, formerly commercial or industrial properties into other types of use.

When evaluating properties with environmentally critical previous use, builders and investors must fundamentally consider that, in addition to potentially costly measures for the remediation of soil, soil air or groundwater, the building fabric of existing operational buildings may still be contaminated with substances such as asbestos, mineral oils, heavy metals, or dioxins, leading to significant additional costs for building demolition and disposal of demolition materials.

Landfill In a landfill, waste is deposited under organized, controlled conditions and thus permanently excluded from the circular economy.

S. Kaub, *Hazardous Materials in the Building Industry*,
https://doi.org/10.1007/978-3-658-49150-5_11

To prevent pollutants from escaping the landfill into the environment, landfills are sealed on all sides, and potential emission pathways (landfill leachate, landfill gas) are monitored.

The requirements for the technical safety of landfills significantly depend on the hazard potential of the waste types stored.

While slightly contaminated construction debris and soil masses can be deposited in mineral waste landfills, toxic waste, such as PCB-containing capacitors or highly contaminated excavated soil, must be disposed of under special conditions (steel drums, encasement) in an underground landfill.

Latency Period The latency period refers to the time interval between the exposure of a human to a health-threatening substance and the appearance of the first symptoms of illness.

Latency periods depend on the substance and can range from hours (e.g. cadmium) to years or even decades (e.g. asbestos).

Leachate Forecasting Leachate forecasts are used to estimate the pollutant inputs from ► contaminated site or areas suspected of contamination into groundwater via leachate.

Possible methods for leachate forecasting include the application of solute transport models, conducting laboratory or in situ investigations, or inferring from downstream measurements.

Lipophilic Hydrocarbons Lipophilic hydrocarbons are an important aggregate parameter in the investigation and assessment of contaminated and potentially contaminated sites.

In the chemical determination of lipophilic hydrocarbons, animal and plant fats and oils, as well as mineral oils and mineral fats, and various aromatic hydrocarbons such as toluene, xylenes, phenols, or naphthalene are included.

M

Man-Made Mineral Fibers Man-made mineral fibers (MMF) are artificially produced glass-like fibers of inorganic origin.

Health risks from MMF primarily arise from the inhalation of microscopically small, respirable fiber components.

The hazard potential of MMF is generally determined through laboratory analysis of the ▶ Carcinogenicity Index.

In the construction industry, MMF products were primarily used as mineral wool insulation materials (glass wool and rock wool) for thermal and sound insulation in building construction.

The recommended protective measures during the remediation of MMF products range from selecting pre-fabricated mineral wool products to low-emission work procedures and personal protective equipment in the form of breathable protective suits or respiratory masks.

S. Kaub, *Hazardous Materials in the Building Industry*,
https://doi.org/10.1007/978-3-658-49150-5_12

Military Contaminated Sites Military contaminated sites refer to ▶ contaminated site that are attributable to the military use of an area after World War II.

Military contaminated sites are found on properties of the German Armed Forces, the former National People's Army (NVA), NATO, and the former Western Group of Soviet Forces (WGT).

The specific environmental damages present in military contaminated sites largely depend on the particular use of the properties.

Typical suspected sites for military contaminated sites are former production facilities, munitions storage sites, defusing locations, blasting and shooting ranges, demilitarization plants, as well as interim and final storage sites for chemical warfare agents.

Military-specific pollutants primarily include mineral oil products (fuels and lubricants), chlorinated hydrocarbons (CHCs) from chemical cleaning, and BTEX aromatics (fuel additives, solvents).

When converting former military properties to other uses, such as residential development, it is important to note that both the land and the building materials of existing structures may be significantly contaminated with environmentally relevant pollutants, leading to increased costs for the remediation of harmful soil changes (contaminated sites) as well as for the demolition and disposal of debris.

Mineral Oil Hydrocarbons Mineral oil hydrocarbons (MOH) are an important aggregate parameter in the investigation and assessment of potentially contaminated sites.

Contaminations by MOH occur particularly at tank facilities of gas stations, airfields, or industrial operations.

The burdens posed by mineral oil hydrocarbons typically affect not only the soil but also frequently impact environmental media such as soil air and groundwater .

While MOH-contaminated soils are generally excavated promptly during construction projects and then remediated or deposited off-site, the remediation of soil air and groundwater damage is often more time-consuming, technically complex, and therefore more expensive.

In the event of identifiable environmental damage from mineral oil hydrocarbons, it is advisable for the property owner to engage a qualified expert as early as possible to minimize subsequent contamination and thus remediation costs.

Mutagen Mutagenic factors are generally those that can lead to genetic changes in humans or animals.

In the construction industry, mutagenic substances are particularly encountered during the remediation of contaminated sites (e.g. benzene) or building pollutants (e.g. PAHs).

O

Old Deposits Decommissioned waste disposal facilities and other properties where waste has been treated, stored, or deposited. Old deposits include former municipal or company-owned, but especially unauthorized, so-called "wild dumps."

Important information about old deposits (pollutant inventory, spatial extent, parcels, etc.) can be obtained by developers through consultation of officially maintained contaminated site registers.

Off-site Methods In the context of contaminated site management, off-site methods refer to all procedures for treating contaminated soils using stationary remediation facilities located away from the site of contamination.

Typical examples of off-site methods include biological or thermal soil remediation or chemical-physical ▶ soil washing.

S. Kaub, *Hazardous Materials in the Building Industry*,
https://doi.org/10.1007/978-3-658-49150-5_13

On-site Methods In the context of contaminated site management, on-site methods refer to all procedures for treating contaminated soils using a mobile or semi-mobile remediation facility established at the site of contamination.

Typical examples of on-site methods include biological soil remediation or chemical-physical ▶ soil washing.

Organic Compounds Organic compounds, with some exceptions (carbonic acid, carbonates, carbides, oxides), are chemical compounds containing the element carbon.

In contaminated site management, hydrocarbon compounds are particularly significant. Typical examples include environmental damage from oil, gasoline or kerosene contamination or contamination from improper operation of chemical cleaning facilities.

Organoleptic Examination In contaminated site management an organoleptic examination refers to the assessment of environmental samples using the senses.

For the construction industry, the organoleptic examination of soil samples is particularly important in the context of contaminated site investigation or when previously unknown contamination is encountered during earthworks.

During the organoleptic soil assessment, characteristics such as color, odor, and homogeneity of soil samples, or any foreign materials like glass, bricks, or slag, are identified.

Based on the organoleptic findings, the parameters to be examined in subsequent chemical analysis can be adjusted to the specific soil conditions.

After the analysis results are available, the organoleptic soil assessment further aids in expert evaluation through analogical reasoning and plausibility checks.

P

PAH Polycyclic Aromatic Hydrocarbons (PAH) is a collective term for a group of several hundred condensed aromatic carbon compounds.

The most important representative and at the same time the key parameter of PAH is the carcinogenic and mutagenic ▶ Benzo-a-pyrene (BaP).

PAHs are formed during the incomplete combustion of organic materials such as fuels, oil, or coke and are ubiquitously present in the environment due to emissions from cars, airplanes, or power plants.

In environmental analysis, the 16 most commonly occurring PAHs are typically determined according to recommendations by the US Environmental Protection Agency (EPA).

S. Kaub, *Hazardous Materials in the Building Industry*,
https://doi.org/10.1007/978-3-658-49150-5_14

In the construction industry, PAHs play an important role, for example, in the remediation of old sites (gasworks, coking plants), the renovation of buildings (tar-containing parquet adhesives), or in determining the disposal route for road debris (distinguishing between tar-containing and bituminous asphalts).

Parameter In environmental analysis, a parameter is understood to be the quantity to be examined for the assessment of a ▶ contamination.

Parameters include not only individual substances such as arsenic, lead or zinc and groups of substances like mineral oils or PCBs but also chemical-physical properties such as pH value, temperature or humidity.

Pathway of Exposure Humans can absorb pollutants into their bodies in various ways. Human toxicology distinguishes the following pathways of exposure:

Inhalation uptake: Absorption by inhalation
Dermal uptake: Absorption through the skin
Oral uptake: Absorption through the mouth.

The issue of pollutant absorption is particularly relevant in the construction industry during the remediation of hazardous waste sites or building contaminants like asbestos or polychlorinated biphenyls (PCBs).

The toxicological assessment of potential pathways of pollutant absorption serves to estimate the risk they pose and thus to establish protective measures for remediation professionals or other affected individuals.

PCB Polychlorinated Biphenyls (PCB) is a collective term for a group of 209 chlorinated aromatic hydrocarbons.

In the chemical analysis of PCB-contaminated environmental samples, usually only six so-called standard congeners are determined, and the total PCB content is calculated by multiplying the sum concentration by a factor of 5.

In the construction industry, PCBs were particularly used as cooling and insulating fluids in transformers and capacitors (closed applications) or as plasticizers in joint sealants and as flame retardants in paints (open systems).

Due to these applications, PCBs play an important role, especially in the remediation of PCB-contaminated buildings and in the disposal of electrical components during demolition and renovation projects.

PCP Abbreviation for the halogenated hydrocarbon compound Pentachlorophenol. Due to its fungicidal effect, PCP was particularly used in the construction industry as a wood preservative.

To assess the need for remediation measures, a graduated approach of dust, material, and indoor air investigations is usually carried out. Depending on the PCP content of the indoor air, blood and urine tests of the users are also considered.

Approved remediation methods include removal, coating or spatial separation of the PCP-containing product.

As a precaution, long-term PCP values of <0.1 micrograms/cbm indoor air should be aimed for.

Permeability Coefficient In hydrogeology, the permeability of rocks for groundwater is described using the permeability coefficient (kf-value).

Permeability coefficients are particularly important when in the context of assessing contaminated sites statements about the horizontal and/or vertical spread of pollutants dissolved in groundwater are required.

The determination of kf-values can be carried out both by examining soil samples in the laboratory and through pumping tests in the field.

PFAS Per- and polyfluorinated alkyl substances (PFAS) are a group of non-naturally occurring substances with more than 10,000 individual substances that are difficult to degrade according to recent estimates.

PFAS consist of carbon chains of varying lengths, in which the hydrogen atoms are partially (polyfluorinated) or completely (perfluorinated) replaced by fluorine atoms.

PFAS suspected sites are particularly locations where firefighting foams have been used, such as firefighting training areas, fire ponds, airports or areas where large fires have been extinguished.

In the construction industry, PFAS play an important role, especially in the remediation of PFAS-contaminated groundwater (▶ contaminated sites).

pH Value The pH value, defined as the negative decimal logarithm of the hydroxide ion concentration, indicates as a dimensionless number how acidic or alkaline an aqueous solution is.

The measurement range extends from pH = 0 (acidic solution) through pH = 7 (neutral solution) to pH = 14 (alkaline solution).

Determining the pH value is particularly important in the construction industry when it comes to assessing the mobilization potential of soil contaminants in the context of site contamination investigations.

Pollutant Pollutants are generally understood to be substances that, due to their chemical-physical properties, are capable of harming human health and other protected goods such as animals, plants, soil or water.

The construction industry is confronted with a wide range of pollutants such as asbestos, PAHs, PCBs (building renovations), or BTEX, chlorinated hydrocarbons, dioxins, heavy metals (contaminated site remediation).

In the context of remediation or demolition activities, handling pollutants (hazardous substances) requires special measures for occupational safety and pollutant disposal.

POP The term "Persistent Organic Pollutants" (POP) refers to halogenated hydrocarbon compounds that are particularly long-lasting and therefore degrade very slowly in the environment.

Examples of POP compounds include insecticides like DDT or dieldrin, plasticizers and flame retardants like PCB and also dioxins.

Preliminary Investigation A preliminary investigation is the second step in the risk assessment of areas suspected of contamination.

Unlike the non-sampling historical investigation, preliminary investigations involve examining environmental media such as soil, soil air, and groundwater at sites of environmentally critical use (suspected areas) through probing, sampling and subsequent chemical analysis.

Preliminary investigations aim to clarify whether there is sufficient suspicion of harmful soil alteration or contamination at a site suspected of contamination.

If suspicions are confirmed, the exact extent of environmental contamination is determined through further ▶ Detailed investigation.

Preliminary investigations are particularly recommended for property owners when, prior to a planned property sale, the presence of pollutants at sites of environmentally critical use and thus the suspicion of contamination needs to be clarified.

Protective and Restrictive Measures Protective and restrictive measures are intended to prevent or mitigate the dangers, significant disadvantages, or significant nuisances posed by ▶contaminated sites to individuals or the public.

Protective and restrictive measures include, in particular, usage restrictions such as preventing access, limiting certain structural uses or prohibiting the extraction of drinking or industrial water.

Protective and restrictive measures are employed as temporary immediate measures for hazard prevention but also as permanent measures when other remediation methods lack proportionality or have low prospects of success.

R

Radon Radon (Rn 222) is a radioactive noble gas and the second leading cause of lung cancer after smoking.

The main source of elevated radon levels in buildings is the soil, while building materials and drinking water play a minor role according to current knowledge.

Radon enters buildings through leaky floor slabs, cracks in masonry or unsealed penetrations of cables or pipes.

For persistent high radon concentrations >100 Becquerel (Bq) per cubic meter of indoor air, authorities recommend measures such as frequent, intensive ventilation or sealing of the aforementioned radon entry paths. For radon levels >1000 Bq per cubic meter of indoor air, structural measures should be completed within three years.

Recycling Recycling refers to the processing of ▶ waste and its subsequent reintegration into the economic cycle as a secondary raw material.

S. Kaub, *Hazardous Materials in the Building Industry*, https://doi.org/10.1007/978-3-658-49150-5_15

The goals of recycling are to reduce waste volumes and conserve natural resources.

In construction projects, mineral mass wastes such as construction debris and road rubble, as well as construction and demolition wood, are particularly processed and used as recycled building materials.

A prerequisite for the recycling of building materials is, in addition to the technical suitability of the RC materials, the compliance with maximum permissible pollutant concentrations.

Remediation In environmental engineering, remediation refers to the application of technical methods to eliminate or reduce pollutant loads with the primary goal of averting risks to human life and health.

In the context of contaminated site remediation, environmental media such as soil, soil air, or groundwater are remediated from environmental pollutants like mineral oils, while building remediation addresses construction pollutants such as asbestos or PCBs.

Remediation is generally required whenever legally prescribed limits in the form of maximum allowable pollutant concentrations are demonstrably exceeded.

Before extensive remediation measures are undertaken, expert consultants develop case-specific remediation plans, which particularly illustrate the cost-benefit ratio for various remediation options.

Remediation Investigation Remediation investigations refer to environmental technical studies aimed at deciding the type and extent of suitable, necessary and appropriate measures for the remediation of a contaminated site.

Remediation investigations can be mandated by the responsible contaminated site authority to the ▶ waste responsible party for remediation in cases where a coordinated approach is necessary due to the diversity of required measures, or where the nature, spread, or quantity of pollutants pose significant harmful soil changes or other dangers to individuals or the public.

Remediation investigations are not required if the dangers, significant disadvantages, or significant nuisances posed by a contaminated site can be countered with simple means or eliminated in other ways (e.g. through natural retention or degradation of pollutants).

Remediation Plan A remediation plan summarizes the technical measures intended for the remediation of a contaminated site in individual cases.

Here, the competent contaminated site authority can require the responsible party for remediation to submit a remediation plan under certain conditions.

This is the case for contaminated sites where a coordinated approach is necessary due to the diversity of required measures or where, due to the nature, spread, or quantity of pollutants other dangers to individuals or the public are particularly significant.

The creation of a remediation plan is not required if the dangers, significant disadvantages, or significant nuisances emanating from a contaminated site can be countered with simple means or eliminated in other ways (e.g. through natural retention or degradation of pollutants).

Responsible Party for Remediation The responsible party for remediation is the natural or legal person legally accountable for remediating a hazard situation caused by environmentally relevant pollutants to humans and/or the environment.

In cases of building contamination by construction pollutants such as asbestos, PCBs, or old mineral fiber products, the responsible party is typically the building owner.

For contamination of environmental media such as soil, soil air, or groundwater, the circle of responsible parties extends to various groups.

Those obligated to remediate a ▶ contaminated site include, as so-called action disruptors, the originator of a harmful soil change or contaminated site and their universal legal successor.

Additionally, the property owner and the holder of actual control (property possessor) are responsible in their capacity as condition disruptors.

If multiple parties are considered responsible for remediation, the competent approval authority has what is known as discretionary selection.

In selecting the responsible party or parties for remediation, criteria such as ensuring effective hazard prevention, the personal and material proximity of the responsible party, their contribution to causing the hazard and the economic capacity of the responsible party play an important role.

Risk Assessment Risk assessment is the second main phase in the processing of contaminated site projects, following the initial survey.

Risk assessments are prepared by expert assessors and serve the responsible contaminated site authority in making the final determination: contaminated site: Yes or No?

In the risk assessment, the findings of all previously conducted investigations, such as historical exploration, preliminary and detailed investigations are summarized and evaluated.

Property owners can be required by the responsible contaminated site authority to conduct the necessary investigations for risk assessment if there is sufficient suspicion of a ▶ contaminated site based on concrete evidence.

S

Safety and Health Protection Coordinator A Safety and Health Protection Coordinator is the person appointed by the client in order to ensure safety and health protection on construction sites.

Only persons with the necessary expertise may act as SAHP-Coordinator.

The coordinator's tasks during the planning phase primarily involve creating a safety and health protection plan and coordinating the monitoring of safety and health protection during the execution of a construction project.

Safety and Health Protection Plan In a Safety and Health Protection Plan the organizational, technical and personal measures for the occupational safety of employees are determined during the planning phase of a construction project based on a hazard analysis.

S. Kaub, *Hazardous Materials in the Building Industry*,
https://doi.org/10.1007/978-3-658-49150-5_16

Safety and health protection plans are created by expert assessors, so-called Safety and Health Protection Coordinators.

Selective Demolition Selective demolition refers to the proper dismantling of structures or parts of structures in various pre-planned stages of dismantling, combined with the separation and separate disposal of the resulting waste.

The selective demolition of commercially or industrially used buildings serves, in addition to the legally required waste separation, to eliminate proven structural burdens, such as oil-contaminated hall floors or the remediation of building pollutants like asbestos, PCBs, heavy metals or dioxins.

The separation and separate disposal of uncontaminated, contaminated, and polluted building materials generally lead to significant savings in demolition costs for the client/investor.

Sick Building Syndrome The term Sick Building Syndrome encompasses various health issues experienced by building occupants, such as irritation of the eyes, nose and upper respiratory tract, fatigue, headaches, nausea, dizziness, or an increased frequency of coughing or respiratory infections.

Office workers in air-conditioned environments are particularly affected by Sick Building Syndrome.

In addition to biological factors like bacteria and fungi from HVAC systems, physical factors such as temperature, humidity and lighting, chemical pollutants in indoor air from cleaning agents, furniture or tobacco smoke, as well as psychological stressors like stress or screen work can contribute to the onset or exacerbation of Sick Building Syndrome.

The difficulties in diagnosing Sick Building Syndrome arise from the multitude of potential causes, the complex interaction mechanisms of the involved pollutants ("pollutant cocktail") and not least the varying sensitivity and resilience of those affected.

Soil Air Soil air refers to the air located in the pore space of the soil above the groundwater level, known as the unsaturated soil zone.

Improper handling of volatile organic solvents, such as chlorinated hydrocarbons (CHCs), has led to particularly in metalworking operations or chemical cleaning facilities in addition to soil and groundwater pollution, often lead to significant contamination of soil air.

The remediation of soil air damages often presents a logistical challenge in construction projects, aside from cost considerations, as the removal of surface sealing is generally only permitted once regulatory remediation targets are met.

Soil Air Remediation Soil air remediation (soil gas extraction) refers to a pneumatic process for cleaning the unsaturated soil zone.

In this remediation technique, volatile contaminants are extracted from the soil using special suction units such as side channel blowers or vacuum pumps through soil air wells (gas wells).

The contaminated soil air is then purified using various methods, such as adsorption with ▶ contaminated carbonsite or oxidation.

Typical applications of soil air extraction include the remediation of chlorinated hydrocarbon (CHC) contamination from former dry-cleaning sites or the remediation of BTEX from decommissioned gas station sites.

Soil air extractions impact construction projects, like all remediation measures, both in terms of cost and site logistics and planning.

The purification of soil air to below regulatory remediation target values is generally a prerequisite for the removal of surface sealing and thus for the execution of excavation work.

Soil Contamination Soil contamination refers to the pollution of soil with harmful substances introduced by humans (anthropogenic) directly or indirectly through air or water.

Anthropogenic soil contamination should be distinguished from naturally occurring, geogenic contamination of soils by "pollutants" such as heavy metals like lead or arsenic.

The early identification of the type and extent of potential soil contamination is of great importance for construction projects.

Through targeted environmental engineering investigations prior to construction activities, necessary remediation measures can be planned in advance and/or potential disposal routes for contaminated excavated soil can be assessed for their compliance.

Soil Function The soil fulfills the following three main functions:

1. Natural functions
   as a basis and habitat for humans, animals, plants, and soil organisms, a component of the natural ecosystem, particularly with its water and nutrient cycles, a medium for the degradation, balancing, and formation of substances due to its filtering, buffering, and

transformation properties, especially for groundwater protection.

2. Functions as an archive of natural and cultural history and
3. Utilization functions as a raw material repository, an area for settlement and recreation, a site for agricultural and forestry use, a site for other economic and public uses, transportation, supply and disposal.

Soil Remediation The term soil remediation encompasses various technical methods for the removal or reduction of soil contaminants.

Soil remediation methods are particularly used in the remediation of ▶ contaminated site but also for the treatment of contaminated soils generated during construction projects.

The selection of the soil remediation method largely depends on the specific contaminants identified and the type of contaminated soil.

Soil Replacement Soil replacement refers to a technical process for the remediation of contaminated soils, as encountered in the context of contaminated site remediation or construction projects.

In the first step, the contaminated soils are excavated by construction machinery under predetermined occupational safety conditions.

The verification of successful excavation is generally conducted through sampling of the walls and base of the excavation pit and chemical analysis of the soil samples for the specific pollutants present.

The excavated soils are then remediated on the construction site (on site) or in specialized soil treatment centers outside the construction site (off site), depending on the identified spectrum of pollutants and the existing soil conditions using various soil remediation methods.

After the completion of soil remediation, the final step involves the reinstallation of the treated excavated materials with permissible residual levels of pollutants.

Soil Washing The term soil washing refers to various washing processes for the remediation of contaminated soils.

Soil washing processes are particularly used for the remediation of ▶ contaminated site but also for contaminated soils generated during construction projects.

Soil washing plants consist of multi-stage classification, washing, and mixing units. The separation of contaminants is achieved through water, possibly with additives, and the application of mechanical energy in the form of stirring, shearing, impact or high pressure.

The cleaned soil masses are, as far as possible, recycled. The separated contaminants remain with the fine-grained fraction as contaminated sludge and are disposed of separately.

The washing of contaminated soils can be carried out either in a mobile plant set up on the construction site (on-site) for the duration of the remediation measure or off-site in stationary soil cleaning centers.

Soil Zone In hydrogeology, the term soil zone refers to rock bodies in solid or loose rock that are completely (water-saturated soil zone) or incompletely (unsaturated soil zone) filled with groundwater at the time of observation.

Various active and passive methods are applied for the remediation of the water-saturated soil zone.

The remediation of the unsaturated soil zone is generally carried out using the method of subsoil-remediation.

Stripping In process engineering, stripping refers to the transfer of a dissolved substance from the liquid phase to the gas phase.

For the construction industry stripping processes are particularly important in the remediation of groundwater contamination caused by legacy pollution.

The use of stripping processes is limited to volatile pollutants such as BTEX aromatics or VOCs and is therefore not applicable to all contaminants.

The remediation of groundwater contamination through stripping is conducted either by injecting compressed air into the contaminated aquifer (in situ stripping) or using so-called stripping towers, where the compressed air is introduced into packed columns countercurrent to the extracted contaminated groundwater.

The pollutant-laden soil air generated during in situ stripping is extracted and, like the exhaust air from the stripping column, subsequently purified through various methods such as adsorption on activated carbon.

Survey The survey serves as the first and fundamental step in the remediation of contaminated sites, involving the systematic collection of data on contaminated areas.

The survey of contaminated areas is also referred to as initial survey, comprehensive survey, inventory or area inventory.

The data obtained during the survey primarily serve to decide whether further investigations are necessary to clarify a suspected contamination.

The results of the survey are recorded in officially maintained contaminated site registers and can be reviewed by developers prior to property purchases or construction activities at the relevant contaminated site authority.

Suspected Contaminated Site A suspected contaminated site is a property where there is suspicion of contamination of environmental media such as soil, soil air, or groundwater.

A suspicion arises when, due to the previous use of a property or specific damage incidents (accidents, war events), there are actual indications of contamination.

The environmental technical investigation of suspected contaminated sites for possible pollutant entries is always required for builders or property owners when the issue of legacy contamination needs to be clarified in advance of construction measures or planned property sales.

T

Temporary Storage Temporary storage refers to the time-limited storage of ▶ waste for the purpose of later disposal.

The temporary storage of waste is particularly significant for the construction industry when contaminated bulk waste such as construction debris or excavated soil with harmful contaminants is generated during construction projects and cannot be directly disposed of for various reasons.

When temporarily storing contaminated construction waste at the site of generation, it must be ensured that the welfare of the public, particularly the protection of groundwater is not compromised.

The specific precautions to be taken depend on the type and extent of contamination and include, for example, impermeable storage areas, the covering of waste to protect against precipitation or the controlled drainage of surface water.

S. Kaub, *Hazardous Materials in the Building Industry*,
https://doi.org/10.1007/978-3-658-49150-5_17

Tetrachloroethylene Tetrachloroethylene (Perchloroethylene, Per) has been primarily used as a technical solvent for cleaning and degreasing. There is a reasonable suspicion that tetrachloroethylene has carcinogenic potential.

Due to its chemical-physical properties (low viscosity, high volatility), improper handling of tetrachloroethylene often led to contamination of soil as well as pollution of environmental media such as soil air and groundwater.

For the construction industry, tetrachloroethylene is particularly significant as a pollutant in the remediation of ▶ contaminated sites such as chemical cleaning facilities or metal processing plants.

Thermal Soil Remediation Thermal soil remediation encompasses all methods for decontaminating polluted soils through the application of high temperatures.

Thermal soil remediation is particularly used for the remediation of contaminated sites but also for cleaning contaminated soils generated during construction projects.

The remediation of contaminated soils through thermal methods is generally conducted in stationary soil treatment centers away from the site of contamination (off-site).

Thermal soil remediation methods are limited to volatile organic pollutants such as solvents or hydrocarbons like BTEX aromatics, PAHs, PCBs, or dioxins and are therefore not applicable to all soil contaminants.

Thermal soil remediation is carried out in three procedural steps.

First, the soil to be remediated is mechanically processed using screening and crushing equipment. In the second step, the contaminated soil masses are heated to temperatures of approximately 500 to 1200 degrees Celsius, converting the volatile pollutants into gas form.

Finally the pollutant-laden exhaust gas is purified through various exhaust gas treatment methods (afterburning, condensation or flue gas cleaning).

The pollutant-free, biologically inert soil can be prepared for intended reuse by adding organic material.

The residues from the exhaust air treatment (filter dusts) are subjected to separate disposal.

Threshold Value In environmental engineering, a threshold value refers to legally binding concentrations of pollutants in a medium as specified in laws, regulations or administrative guidelines.

Important threshold values in the construction industry include those for assessing soil quality, classifying excavated soil or construction debris, or the maximum permissible indoor air concentrations as proof of successful remediation after building renovation.

TNT Trinitrotoluene (TNT) belongs to the class of ▶ aromatic hydrocarbons and is the most significant explosive used for civilian and military purposes.

There is a reasonable suspicion that TNT has carcinogenic potential.

TNT and other nitroaromatic compounds are particularly significant for the construction industry in the context of remediating ▶ military sites, such as explosive factories or detonation sites.

Trichloroethylene Trichloroethylene (Tri) has been primarily used as a technical solvent for cleaning and degreasing.

There is a reasonable suspicion that tetrachloroethylene has carcinogenic potential.

Due to its chemical-physical properties (low viscosity, high volatility), improper handling of Trichloroethene often leads to contamination of environmental media such as soil air and groundwater in addition to soil contamination.

For the construction industry, trichloroethene is particularly significant as a pollutant in the remediation of ▶ contaminated sites such as dry-cleaning facilities or metal processing operations.

U

Ubiquitous In environmental engineering, "ubiquitous" (Latin: ubique = everywhere) refers to the distribution of pollutants across all environmental media.

Pollutants that are ubiquitously detectable include, for example, dioxins and PAHs (emissions from motor vehicles, aircraft, power plants), asbestos (wear from roofing materials) or PCBs (manufacturing and use).

Underground Landfill By depositing waste in an underground landfill (UGL), environmentally hazardous waste is permanently excluded from the circular economy.

Waste is typically stored in already mined salt mines, as these geological formations are considered particularly safe.

Specific requirements are imposed by landfill operators on the type of waste stored. For example, UGL waste must not be explosive, self-igniting, or gas-forming under storage conditions.

S. Kaub, *Hazardous Materials in the Building Industry*, https://doi.org/10.1007/978-3-658-49150-5_18

UGL waste generated during construction projects includes, for example, contaminated excavated soil, PCB-containing capacitors or broken fluorescent tubes.

Waste producers (builders, contractors) must adhere to the storage conditions set by the landfill operator (e.g. steel drums or containers) when disposing of waste in an underground landfill.

Utilization Waste can be disposed of either through elimination (e.g. by landfilling) or through utilization.

In waste recycling, a distinction is generally made between material recycling and energy recovery.

While material recycling involves obtaining (secondary) raw materials from waste, energy recovery includes the use of waste as a substitute fuel for energy generation.

For producers of construction waste, the priority of recycling over waste disposal is initially significant, as long as it is technically feasible and economically reasonable.

In construction projects, it is generally important to separate the waste generated, such as wood waste, metals, roofing felt, construction debris and asbestos products, by fractions and to subject them to separate disposal.

V

VOC Volatile Organic Compounds (VOC) are defined by the WHO as a group of volatile organic compounds with a boiling range from 50–100°C to 240–260°C.

Possible VOC sources include outdoor air and vehicle traffic, as well as bitumen sheets and tar adhesives, but also household cleaners, solvents and disinfectants.

Volatile organic compounds in the elution range between n-hexane and n-hexadecane are collectively referred to as TVOC (Total Volatile Organic Compounds).

The TVOC content of indoor air is used to assess the indoor air hygiene situation of interior spaces.

Volatile Halogenated Hydrocarbons Volatile Halogenated Hydrocarbons (VHH) are hydrocarbons in which one or more hydrogen atoms have been replaced by the elements fluorine, chlorine, bromine or iodine (halogens).

S. Kaub, *Hazardous Materials in the Building Industry*,
https://doi.org/10.1007/978-3-658-49150-5_19

Among the most well-known representatives of VHH are the compounds trichloroethylene (Tri) and tetrachloroethylene (Per), which were used as solvents, particularly in chemical cleaning and metalworking industries.

In the context of the remediation of ▶ contaminated site, VHH often appear as contaminants in environmental media such as soil, soil air and groundwater.

The remediation of VHH-related environmental damage is often prolonged and therefore costly, especially in cases of groundwater contamination.

W

Waste Waste generally refers to all substances or objects that their owner discards, intends to discard, or is required to discard.

Depending on the disposal route, a fundamental distinction is made between waste for recovery and waste for disposal.

Waste Code Waste codes are multi-digit numerical codes for the clear classification of different types of waste.

In the member states of the European Union, waste is classified according to its origin and composition using 6-digit waste codes into 20 main groups and corresponding subgroups.

Waste for Disposal Waste for disposal is deposited in approved landfills or permanently excluded from the circular economy through chemical-physical or thermal treatment.

S. Kaub, *Hazardous Materials in the Building Industry*,
https://doi.org/10.1007/978-3-658-49150-5_20

In the context of construction projects, waste for disposal includes, for example, asbestos products, heavily contaminated old wood or construction debris and excavated soil with harmful contaminants.

Waste for Recovery The recovery of waste can generally be for the production of new products (material recovery) or for energy generation (energy recovery).

In the context of construction projects, the predominantly occurring, non- or slightly contaminated mineral mass waste such as construction debris and excavated soil is mostly used in earthworks, road construction, landscape construction and landfill construction or for backfilling construction pits or reclamation measures, thus being subjected to material recovery.

Waste Owner Any natural or legal person who has actual control over waste.

The basic obligations of waste owners (construction contractors) include, in particular, the duty to keep construction waste separate, the duty to recover construction waste as far as technically possible and economically reasonable and the duty to dispose of construction waste in a manner compatible with the public interest.

Waste Producer Any natural or legal person whose activities produce waste (primary producer) or who carries out pre-treatment, mixing or other operations that result in a change in the nature or composition of this waste (secondary producer).

The basic obligations of waste producers (client, construction contractor) include, in particular, the duty to keep construction waste separate, the duty to recover construction waste as far as technically possible and economically reasonable and the duty to dispose of construction waste in a manner compatible with the public interest.

Waste Wood Waste wood refers to both wood residues from wood processing and manufacturing industries (industrial residual wood) and used products made from solid wood, wood-based materials or composites with a predominant wood content (used wood), as long as these are classified as waste.

The disposal of waste wood can generally be carried out through material or energy recovery or through thermal disposal.

In construction projects, waste wood may arise as formwork, pallets, cable drums, roof structures, cladding, doors, window frames, furniture, or parquet. Depending on previous use, construction and demolition wood may be treated with a variety of substances such as paints, varnishes, adhesives or wood preservatives and may require special disposal as treated, contaminated or particularly contaminated waste wood.

White Area A white area is defined as the area not contaminated by pollutants.

In the white area, for example are the rest and changing rooms for remediation specialists or the offices of the supervising construction management.

The white area is typically separated from the remediation area ("black area") by special airlocks to prevent the transfer of respective pollutants.

WHO Fibers WHO fibers, as defined by the World Health Organization (WHO), are fibers with critical geometric dimensions (fiber length $L > 5\ \mu m$, fiber diameter $D < 3\ \mu m$, ratio $L : D > 3$).

The construction industry is particularly affected by WHO fibers during demolition projects and the necessary removal of old mineral wool insulation materials used for thermal and sound insulation.

Wood Preservatives Wood preservatives are substances used in the treatment and processing of wood with biocidal effects against wood-destroying insects or fungi, as well as wood-discoloring fungi and substances to reduce the flammability of wood.

Wood preservatives are broadly categorized into inorganic salt compounds and organic wood preservatives such as PCP, lindane or tar oils.

The use of wood preservatives indoors can seriously endanger human health due to increased indoor air pollution. Depending on the specific circumstances, remediation of the affected rooms or buildings may be necessary.

The use of wood preservatives also leads to increased disposal costs for contaminated ▶ waste wood generated during remediation or demolition projects.

Work in Contaminated Areas Work in contaminated areas refers to all activities in areas containing hazardous substances in amounts harmful to humans and/or the environment.

In the construction industry this particularly includes the demolition of facilities and buildings contaminated with pollutants, the remediation of contaminated sites or fire damage or activities in landfill construction.

Remediation sites typically consist of the actual remediation area (black area) and a publicly accessible area (white area) separated from it by special airlocks.

The proof of remediation success is provided, depending on the present pollutants, either through visual inspection, by taking contact, wipe or material samples or by technical examination of the indoor air.

References

Bickel, C.: Bundes-Bodenschutzgesetz – Kommentar Carl Heymanns Verlag KG Köln, Berlin, Bonn, München (2000)

Burmeier, D., et al.: Sicheres Arbeiten auf Altlasten – Handlungsanleitungen für Aufsichtsbehörden, Planer und Ausführende

Gesetz zur Förderung: der Kreislaufwirtschaft und Sicherung der umweltverträglichen Bewirtschaftung von Abfällen (Kreislaufwirtschaftsgesetz – KrWG). Zugegriffen: 24. Febr. 2012.

Gesetz zum Schutz: vor schädlichen Bodenveränderungen und zur Sanierung von Altlasten (Bundes-Bodenschutzgesetz – BBodSchG). Zugegriffne: 17. März (1998)

Gihr, R., Daniel, B., et al.: Altlastenanalytik – Parameter-liste zur branchenspezifischen Auswahl von Analysepara-metern für Altstandorte. Landsberg (1990)

Göbel, W.: Gefahrstoff-ABC. Landsberg (1991)

Hempfling, R., Steffen, S.: (Hrsg.) Schadstoffe in Gebäuden – Erkennen, Bewerten, Sanieren, Vermeiden. Taunusstein (1994)

S. Kaub, *Hazardous materials in the building industry*, https://doi.org/10.1007/978-3-658-49150-5

Hölting, B.: Hydrogeologie – Einführung in die All-gemeine und Angewandte Hydrogeologie. Stuttgart (1992)

Hoppe, W., Beckmann, M., Kauch, P.: Umweltrecht. München (2000)

Knopp, L., Eike, A.: Altlastenrecht in der Praxis Unter Berücksichtigung des Bundes-Bodenschutzgesetzes. Herne (1992)

Koch, R.: Umweltchemikalien – Physikalisch-chemische Daten, Toxizitäten, Grenz- und Richtwerte Umweltver-halten Weinheim (1995)

Landesanstalt für Umweltschutz Baden-Württemberg.: Handbuch Bodenwäsche. Stuttgart (1992)

Landesanstalt für Umweltschutz Baden-Württemberg.: Handbuch Altlasten – Genehmigungserfordernisse bei der Altlastbearbeitung. Stuttgart (1995)

Müller, V.: Sonderabfallentsorgung in der Bundesrepublik Deutschland. Hannover (1994)

Neumaier, H., Weber, H. (Hrsg.).: Altlasten – Erkennen, Bewerten, Sanieren. Berlin (1996)

Schultheiß, S., Goos, W.: Altlasten – Eine Einführung für Naturwissenschaftler, Ingenieure und Planer. Köln (1993)

Technische Regel für Gefahrstoffe: TRGS 519 Asbest – Abbruch-, Sanierungs und Instandhaltungsarbeiten Ausgabe Januar (2014)

Technische Regel für Gefahrstoffe: TRGS 521 Abbruch-, Sanierungs und Instandhaltungsarbeiten mit alter Mineralwolle, Ausgabe Februar (2008)

Umweltbundesamt.: PFAS. Gekommen, um zu bleiben. Schwerpunkt 1. Dessau (2020)

Umweltbundesamt.: Sanierungsmanagement für lokale und flächenhafte PFAS-Kontaminationen, Abschlussbericht. Dessau (2020)

UVP Förderverein (Hrsg.).: Umweltverträglichkeitsprüfung – Leitfaden für Unternehmer. Hamm (1990)

Verordnung zum Schutz vor Gefahrstoffen.: (Gefahrstoff-Verordnung – GefStoffV). Zugegriffen: 23. Dez. 2004

Zentralverband des Deutschen Baugewerbes. (Hrsg.).: Umweltgerechter und kostensparender Umgang mit Bauabfällen. Bonn (1997)